21st
Century
Skills Library

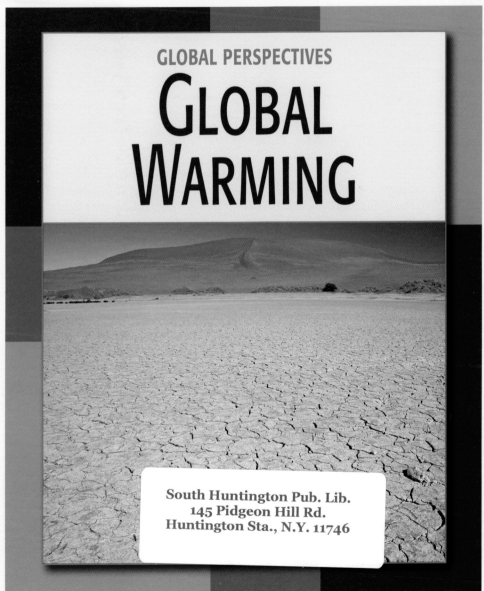

GLOBAL PERSPECTIVES

GLOBAL WARMING

Robert Green

Cherry Lake Publishing
Ann Arbor, Michigan

J 363.738
Green

CHERRY LAKE
Publishing

Published in the United States of America by Cherry Lake Publishing
Ann Arbor, Michigan
www.cherrylakepublishing.com

Content Adviser: Dr. Kevin Trenberth, Head, Climate Analysis Section, National Center for Atmospheric Research, Boulder, Colorado

Photo Credits: Cover and page 1, © iStockphoto.com/ra-photos; page 4, © Jon Arnold Images/Alamy; page 6, © Mayskyphoto, used under license from Shutterstock, Inc.; page 8, © Nick Cobbing/Alamy; page 10, © Bryan Busovicki, used under license from Shutterstock, Inc.; page 12, © Romeo Koitmäe, used under license from Shutterstock, Inc.; page 13, © David R. Frazier Photolibrary, Inc./Alamy; page 14, © Images of Africa Photobank/Alamy; page 17, © WorldFoto/Alamy; page 18, © Elisei Shafer, used under license from Shutterstock, Inc.; page 19, © egd, used under license from Shutterstock, Inc.; page 20, © Wolfgang Kaehler/Alamy; page 21, © Ashley Cooper/Alamy; page 23, © Jupiter Images/Brand X/Alamy; page 25, © David Robertson/Alamy

Map by XNR Productions Inc.

Library of Congress Cataloging-in-Publication Data
Green, Robert, 1969–
 Global warming / by Robert Green.
    p. cm.—(Global perspectives)
 Includes index.
 ISBN-13: 978-1-60279-123-7
 ISBN-10: 1-60279-123-6
 1. Global warming—Juvenile literature. I. Title. II. Series.
 QC981.8.G56G726 2008
 363.738'74—dc22   2007033719

Cherry Lake Publishing would like to acknowledge the work of
The Partnership for 21st Century Skills.
Please visit www.21stcenturyskills.org for more information.

TABLE OF CONTENTS

CHAPTER ONE

A CHILLY SUMMIT

*The city of Interlaken, Switzerland, is well known
for the beauty of its snowy mountain peaks.*

Muhammad al-Hakim, still weary after his flight from Egypt, stood
at the window of the convention center in the town of Interlaken,

Switzerland. He was one of a group of student leaders from around the world who had been selected to participate in workshops on various global issues. His group was studying global warming.

"I've never seen so much snow and ice," he said to a girl who wandered over to share the view. "Egypt is a desert country, with few mountains and very little water."

Standing beside Muhammad was a serious girl named Marie Pepin, who came from Canada. "If global warming continues," she said, "the mountains will have less and less snow on them over time."

Muhammad, feeling a slight chill, didn't know if it would be such a bad thing if Switzerland were a little warmer in the winter. Another student, Tommy Tuttle, from the U.S. city of Boston, Massachusetts, spoke up with his own doubts. "You know, it's not proven that global warming is a permanent part of life on Earth," he said. "It might just be that we are going through a warm cycle in Earth's climate."

Marie looked a bit confused. "Don't most scientists agree that humans contribute to global warming, even if Earth's climate heats up and cools off in cycles?" she asked.

Tommy had enjoyed the plane ride from the United States to Switzerland. He hopes one day to become a pilot, he said, "because lots of people travel by plane, so the world will need lots of pilots."

Some scientists concerned about the environment study the effect of aircraft emissions on the atmosphere.

"But do you know," said Marie, "that air travel may contribute to global warming? The fuel burned by planes releases gases called **emissions**, which may be harmful to the **atmosphere**."

Tommy and Muhammad didn't like hearing this. They decided that they needed to find out whether Marie was correct.

✳ ✳ ✳

What is global warming? Generally, "warming" refers to making things hotter. Global warming means the heating of the Earth. Many people believe that in the past 50 years, most of the planet's warming has been caused by human activities. These activities are said to change the composition of the atmosphere. The main problem is the burning of fossil fuels, which adds carbon dioxide to the atmosphere. One consequence of this heating is increased temperatures.

Some people think of this increase in temperature as global warming, but global warming is really much more. The extra heat causes increased drying, or **evaporation**. Increased evaporation may result in more droughts in places where there is already little rain. It may also lead to heavier rains that may cause flooding in areas where it rains.

21st Century Content

In democratic countries, leaders are elected. These elected government officials have an obligation to work for the interests of the people they represent. One common problem for these elected officials is to decide what is best in the short run and what is best in the long run. If they help people now, they might actually be harming those same people's long-term interests. But politicians want to be re-elected, so they tend to look for quick solutions.

For example, if the government requires measures that reduce the use of electricity, it might cause prices for some things to go up. The measure would therefore be unpopular, and politicians might avoid it. But the measure might help people in the long term because it would reduce pollution from the coal that is burned to create the electricity. There are no easy answers to these problems. To address global warming, leaders will need to balance short-term and long-term needs.

CHAPTER TWO

LIVING IN A GREENHOUSE

A scientist adjusts a camera used to monitor a glacier in Greenland.

The students first addressed the question, what is global warming? One viewpoint that they found particularly interesting was that of Helene Hansen, who was from the island nation of Greenland. "My country today is warmer than it has been for as long as anyone can remember," said

Helene, "and life is changing. This is a very real concern for the people of Greenland."

❋ ❋ ❋

Global warming itself is already an accepted fact. What remains in question is whether humans are the cause of it. Scientists agree that global air and sea temperatures have been rising for the past 100 years. They also know that these temperatures have risen and fallen regularly throughout geologic history. This is known as global climate change. Natural global climate change takes place slowly, over hundreds of years. According to many scientists, however, for the last 50 years temperature increases have been occurring 100 times faster than natural climate change.

There are many things that cause the climate to change. Some of them occur naturally and some of them may be caused by human activity. Scientists understand that the global warming that occurred in the first half of the 20th century

How do we know that Earth's climate is changing? Scientists collect data by measuring many different things. They analyze temperature measurements taken with thermometers at tens of thousands of sites throughout the world. These sites include ships at sea and balloons sent into Earth's atmosphere. Scientists use instruments called bathythermographs to take temperature measurements of Earth's oceans. They also measure snowfall in specific places over time. They measure the size of glaciers and the amount of water vapor in the air. They also observe rising sea levels.

If you were a scientist gathering data to help you understand how Earth's climate is changing, what other kinds of information would you want to collect?

Volcanic gases rise from the crater of Kilauea in Hawaii.
Volcanic eruptions have an effect on Earth's climate.

was partly caused by volcanic eruptions and by changes in the amount of energy given off by the sun, known as **solar radiation**. Volcanic eruptions inject gases and particles into the stratosphere (the part of Earth's atmosphere above where weather occurs and where many jet aircraft fly). There the particles block the sun and cause cooling. The amount of solar radiation also varies. During some periods, the sun heats Earth's atmosphere faster than at other times, but the effect is considered to be small.

These factors are natural. The main cause of natural climate variations is the regular changes in Earth's orbit around the sun. This causes the ice ages and interglacials (warm periods between ice ages) over periods of tens of thousands of years.

Another cause of global warming is the changing composition of the atmosphere. Human life on Earth probably could not continue without the trapping of heat by specific gases in the atmosphere. These gases keep the temperatures just right, making life on the planet possible. It's not too hot or too cold. As early as 1827, a French scientist named Joseph Fourier wrote that gases in the atmosphere trap heat and make Earth's surface warmer.

It is known, however, that the release of certain gases by humans or by natural processes is changing the atmosphere's composition. **Greenhouse gases** released from Earth stay in the atmosphere a long time. They trap heat near Earth's surface, which causes temperatures to rise. This is known as the **greenhouse effect**, because the result is similar to what happens in a greenhouse. Greenhouses are made of glass. The glass traps the sun's rays, causing the inside of the greenhouse to warm up because the heat does not escape from the greenhouse.

Some greenhouse gases are released naturally. An example is methane, a gas that is released from marshes and swamps. Another example is carbon dioxide, a gas that comes from many sources, including when you exhale, or breathe out. The burning of **fossil fuels** also releases large amounts of carbon dioxide. Other greenhouse gases that may be building up because of human activities are nitrous oxide and ozone. The amount of greenhouse

Rotting plants in swamps create methane, which is released into the air.

gases building up in the atmosphere is increasing, and this is speeding up the warming of Earth's climate. Part of the global warming debate is what should be done about the release of these gases, which build up easily in the atmosphere because they don't break down quickly.

CHAPTER THREE

THE DEBATE HEATS UP

*Offshore oil rigs, such as this one in the South China
Sea, produce a fossil fuel known as petroleum.*

"This is where your airplane comes in," Marie said to Tommy. "The
fuel it burns is a fossil fuel, which when burned, releases carbon dioxide
into the atmosphere."

Trees have been cut down and burned to clear this land in Kenya for farming.

"Well, what exactly are fossil fuels?" asked Tommy.

✳ ✳ ✳

Fossil fuels are the main sources of energy in modern countries. Fossil fuels include gasoline, oil, natural gas, and coal. These fuels are called fossil fuels because they come from layers of earth where the fossils, or remains, of living things accumulated. The remains of these once-living things, including plants and animals, are changed by the pressure of the layers of earth above them, as well as by chemical processes. After millions of years, these deposits can be mined or tapped for coal, oil, and natural gas.

Humans have been changing the natural environment—mining, clearing land, planting crops—since earliest history. Cutting down trees may actually increase the amount of carbon dioxide in the atmosphere. Trees produce oxygen, which humans and animals need to breathe, and they absorb carbon dioxide. When they decay or are burned, carbon dioxide is released into the atmosphere. In some places, deforestation, or the cutting down of forests, occurred rapidly in the last 200 years, to make room for cities and industries. Many people believe deforestation has contributed to global warming, mainly through the burning of the trees.

As cities grew and industries increased, people began to burn fossil fuels more quickly to generate more energy. Automobiles may be one of the most common sources of greenhouse gases. The factories

Learning & Innovation Skills

The burning of fossil fuels produces greenhouse gases that warm the atmosphere. How do we find alternatives to fossil fuels? This requires the creation of new technologies through a process called innovation. The cars, planes, and machines that are powered by fossil fuels are examples of earlier innovations. Can you imagine new technologies that might help reduce greenhouse gases?

that burn coal also send carbon dioxide into the atmosphere. Carbon dioxide concentrations in the atmosphere have increased by more than 35 percent since the mid-1700s. Based on this and other evidence, most scientsts agree that human activity has contributed to the warming of Earth's atmosphere, especially in the last 50 years.

Climate changes are measured in various ways. One of the best sources of information on global warming may be found in the glaciers in places such as Greenland, where 85 percent of the land is covered by ice. By drilling deep into the layers of ice, scientists can measure the amount of greenhouse gases in past ages and compare the findings to present levels. Scientists can also measure how quickly glaciers are melting. They do this by taking readings of the air and water temperatures, and by measuring how fast a glacier retreats, or moves backward, as its ice breaks off into the ocean. As glaciers warm, they may retreat faster.

Greenland's Jakobshavn Glacier was first photographed in 1850. Over the years, the glacier retreated slowly, then stopped for about 50 years. In 1997, the glacier began to retreat again, and its retreat was twice as fast as it had been previously. It has increased the rate at which sea level rises by about 0.002 inches (0.06 millimeters) per year. Much of this increase in water level, though, is said to be meaningless because the water is lost in evaporation. Some scientists believe the glacier's quicker retreat is the

*Melting water pours off an iceberg that broke
away from the Jakobshavn Glacier.*

result of an increase in Greenland's average temperature, which has risen
by 4 degrees Fahrenheit (2.3 degrees Celsius) over the past decade.

The United Nations set up the Intergovernmental Panel on Climate
Change (IPCC) to release findings on climate change and to assess
scientific findings and the possible role of humans. A 2007 IPCC report
on the findings of some of the world's scientists concluded that human
activity is likely causing global warming.

People worried about global warming fear that animals won't be able to adapt to a more rapidly changing climate.

Some climate scientists, however, disagree with these findings. They place more emphasis on the theory that global warming is a natural **phenomenon**. They point to the evidence that supports a natural cycle of climate change that takes place approximately every 1,500 years. They do not completely trust computer climate models. They say that while the models are useful, they are too simplified to make accurate conclusions about the future of Earth's complex climate.

CHAPTER FOUR

POSSIBLE CONSEQUENCES OF GLOBAL WARMING

Greenhouse gases are released into the air when gasoline is used to power automobiles.

Muhammad and Tommy became very interested at this point. "Well, if there is evidence that global warming is natural, what does it matter if Earth gets a little warmer?" asked Muhammad, still marveling at the chill in the Swiss air.

Fishers in Greenland have been affected by the warming of the oceans that surround their country.

"In Greenland," said Helene, "life is already changing. The warming of the water has changed the kinds of fish found around Greenland, and this means a big change for our local fishers. As the water warms, cold water fish swim toward colder waters, but other fish have begun to show up around our coastal waters."

She explained that this was not necessarily a bad thing. The environment has constantly evolved over time. But the real problem is the rate of change, which some scientists claim is much faster than occurs naturally.

"But aren't you glad to have warmer weather?" asked Tommy.

"Yes," said Helene, "but what if the ice keeps melting? Parts of Greenland could disappear under the water."

✳ ✳ ✳

Because of the melting of polar ice caps and glaciers, some scientists worry that water levels could rise in the coming years. The global sea level is already rising because of melting glaciers and the expansion of the ocean water as it heats up. As the oceans rise, the entire globe could be affected. Some coastal areas could find themselves under water. This is especially

The island nation of Tuvalu could become uninhabitable if ocean levels continue to rise.

21st Century Content

Industrial activity is more concentrated in some places on the globe. The United States, Europe, Japan, and increasingly China and India are all places with a lot of industrial activity. They are the main producers of greenhouse gases. Industrial activity provides jobs and the things we need, but it also creates pollution and can damage the environment. The damage is not limited to the areas with industrial activity. Greenland, for example, may face severe problems caused by global warming, but the country itself produces little pollution and carbon dioxide. Countries are affected by the activities of other countries, and that is why Earth's warming is considered a global issue.

worrisome because coastal areas tend to be more populated than other areas. And they are most vulnerable when a **storm surge** raises sea levels even more.

It is true that in some colder regions, global warming might have some benefits. These benefits include lengthening the amount of time during the year that crops can be grown. But in hotter areas, such as North Africa, deserts are expanding. As the surface temperatures of the land heat up, moisture is sucked out of the land and into the air through evaporation. That water vapor is one of the major contributors to the greenhouse effect. Evaporation may further speed up global warming and dry out areas that have been used to grow food.

The land is not all that is heating up. The oceans are also getting warmer. Warming water may have begun to cause changes in the places where certain fish live. Other problems might arise, as well. As oceans heat up, tropical storms

Was Hurricane Katrina such a strong storm because of global warming? Some scientists think so, but others disagree.

and hurricanes could become more intense, and patterns of rainfall could change. Some scientists who study Earth's climate believe that Hurricane Katrina, which devastated portions of Louisiana and Mississippi in 2005, was made worse by global warming. Other scientists strongly disagree.

It is hard to prove that specific weather events are caused by global warming. A hurricane, for example, is a weather event known to depend

on warm ocean waters for its formation. We know that Earth's oceans are getting warmer. But when a storm leaves a coastal area heavily damaged, no one can say for sure that the storm was made stronger because of global warming. It might just have been a very strong storm, like many others in history.

Finding solutions to the problem of global warming is complicated. There are economic factors to consider. If humans begin to reduce emissions by regulating industrial activity, prices for products could go up. When prices are high, demand for products goes down. In order to stay in business, companies have to reduce their workforces. When people are out of work, there is even less demand for products. This is bad for the economy. But if industries become more energy efficient and produce less waste, then emissions can be reduced.

CHAPTER FIVE

DO HUMANS REALLY CAUSE GLOBAL WARMING?

Greenhouse gases are produced by many different industries. Many world leaders met in Kyoto, Japan, to discuss how to reduce greenhouse gas emissions.

"**B**ut what is the proof that humans are actually causing global warming?" asked Tommy.

"Scientists look at computer models and interpret them to mean that humans are responsible," Marie said. "Leaders in many countries are taking those models seriously. At a meeting in Japan, the leaders of many nations signed an agreement to try to limit greenhouse gases in an attempt to reverse global warming."

＊ ＊ ＊

The Kyoto Protocol was signed in Kyoto, Japan, in 1997. The agreement outlines a plan of action to address climate change. It divides countries into two groups: developed industrial countries and developing countries. Developed countries, which generally produce more greenhouse gases, would try to decrease their output to 5 percent below 1990 levels. Developing countries would be allowed to produce more greenhouse gases than developed nations.

One of the more interesting ideas proposed in the agreement is the idea of trading rights to emit greenhouse gases. Countries can buy other countries' "credits" to produce gases or to pay for projects to reduce greenhouse gases in other countries. In other words, the goal is a global reduction, but countries can work together to do this in a way that suits their interests.

Although more than 175 countries have signed the agreement, the leaders of the United States and Australia have not signed it. They are concerned about the slowdown in economic growth that could result. They also argue that it is unfair that the agreement allows some countries, such as China and India, to produce more greenhouse gases than developed nations.

＊ ＊ ＊

"But what if world leaders don't agree about the cause of global warming? What, if anything, should be done about it? What can we do?" asked Helene, still worried about the melting of Greenland's glaciers.

"We can continue to study both sides of the issue," said Marie. "And even though we don't know for sure that humans cause global warming, we can still take steps to help, just in case. For example, since so much pollution comes from using electricity, we can use energy-saving lightbulbs, make sure we turn off the lights when we leave a room, and use less air-conditioning."

"There are also alternatives to the energy that we use now," said Muhammad, who had learned a lot. "For example, we can use solar power, which creates electricity through the sun's power, to heat our homes. Or we can use the power of the wind to make electricity instead of burning coal or natural gas."

"I wonder," said Tommy, "if I can fly a **glider** instead of a jet? That way, I could be a pilot and not worry about whether I'm contributing to global warming."

Today, English is an international language. This means that when a Japanese pilot lands his plane in China, the pilot speaks English to the Chinese control tower. But not everyone speaks English. At international conferences, delegates often gather in small groups of those who speak the same language. This reflects the importance of learning foreign languages. If issues such as global warming are to be studied through international cooperation, people must be able to communicate by speaking each other's languages. The United Nations, for example, uses six official languages when it conducts business—Arabic, Chinese, English, French, Russian, and Spanish.

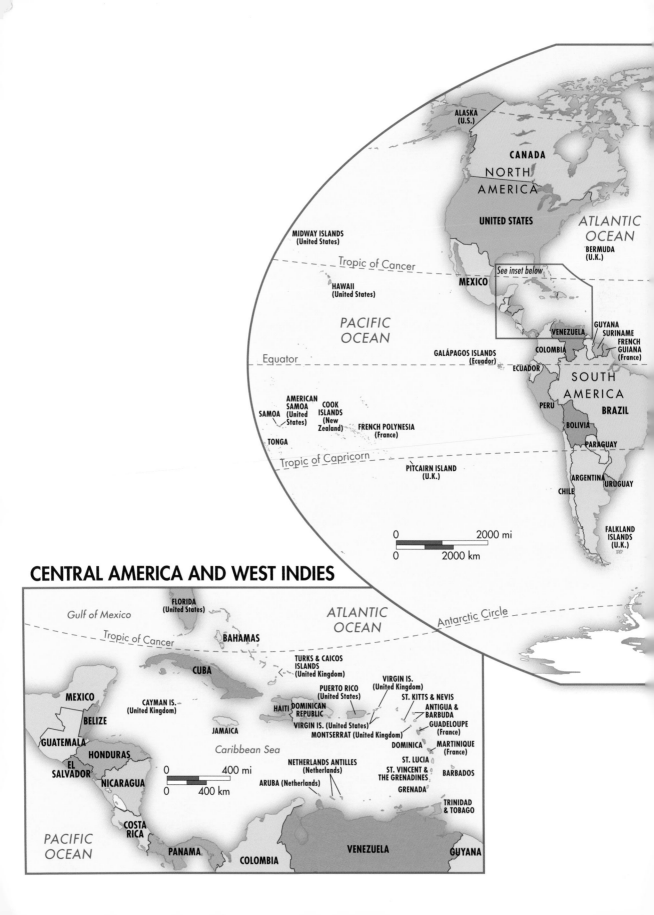

ALASKA
(U.S.)

CANADA

NORTH
AMERICA

UNITED STATES

ATLANTIC
OCEAN

BERMUDA
(U.K.)

MIDWAY ISLANDS
(United States)

Tropic of Cancer

See inset below

MEXICO

HAWAII
(United States)

PACIFIC
OCEAN

GALÁPAGOS ISLANDS
(Ecuador)

VENEZUELA

GUYANA
SURINAME
FRENCH
GUIANA
(France)

COLOMBIA

ECUADOR

Equator

SOUTH
AMERICA

PERU

BRAZIL

AMERICAN
SAMOA
(United
States)

COOK
ISLANDS
(New
Zealand)

SAMOA

FRENCH POLYNESIA
(France)

BOLIVIA

TONGA

PARAGUAY

Tropic of Capricorn

PITCAIRN ISLAND
(U.K.)

ARGENTINA

URUGUAY

CHILE

0   2000 mi

FALKLAND
ISLANDS
(U.K.)

0   2000 km

CENTRAL AMERICA AND WEST INDIES

Gulf of Mexico

FLORIDA
(United States)

ATLANTIC
OCEAN

Antarctic Circle

Tropic of Cancer

BAHAMAS

TURKS & CAICOS
ISLANDS
(United Kingdom)

CUBA

VIRGIN IS.
(United Kingdom)

MEXICO

CAYMAN IS.
(United Kingdom)

PUERTO RICO
(United States)

ST. KITTS & NEVIS

HAITI

DOMINICAN
REPUBLIC

ANTIGUA &
BARBUDA

BELIZE

JAMAICA

VIRGIN IS. (United States)

GUADELOUPE
(France)

GUATEMALA

MONTSERRAT (United Kingdom)

MARTINIQUE
(France)

HONDURAS

Caribbean Sea

DOMINICA

EL
SALVADOR

0   400 mi

NETHERLANDS ANTILLES
(Netherlands)

ST. LUCIA

NICARAGUA

0   400 km

ST. VINCENT &
THE GRENADINES

BARBADOS

ARUBA (Netherlands)

GRENADA

TRINIDAD
& TOBAGO

PACIFIC
OCEAN

COSTA
RICA

PANAMA

COLOMBIA

VENEZUELA

GUYANA

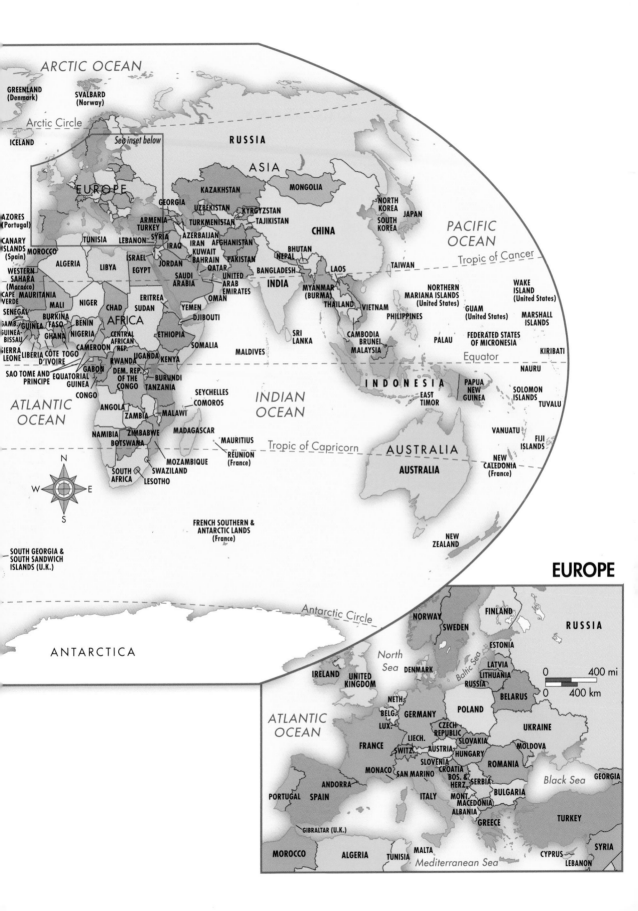

ARCTIC OCEAN

GREENLAND
(Denmark)

SVALBARD
(Norway)

Arctic Circle

ICELAND

RUSSIA

See inset below

EUROPE

ASIA

AZORES
(Portugal)

GEORGIA

KAZAKHSTAN

MONGOLIA

CANARY
ISLANDS
(Spain)

MOROCCO

ARMENIA
TURKEY

UZBEKISTAN

KYRGYZSTAN

TURKMENISTAN

TAJIKISTAN

NORTH
KOREA

JAPAN

SOUTH
KOREA

PACIFIC
OCEAN

TUNISIA

LEBANON

SYRIA

AZERBAIJAN

IRAN
AFGHANISTAN

CHINA

ALGERIA

LIBYA

IRAQ

KUWAIT

PAKISTAN

BHUTAN
NEPAL

TAIWAN

Tropic of Cancer

WAKE
ISLAND
(United States)

ISRAEL

EGYPT

JORDAN

BAHRAIN

QATAR

WESTERN
SAHARA
(Morocco)

SAUDI
ARABIA

UNITED
ARAB
EMIRATES

BANGLADESH

INDIA

LAOS

NORTHERN
MARIANA ISLANDS
(United States)

GUAM
(United States)

MARSHALL
ISLANDS

CAPE
VERDE

MAURITANIA

MALI

NIGER

CHAD

OMAN

YEMEN

SUDAN

MYANMAR
(BURMA)

THAILAND

VIETNAM

PHILIPPINES

SENEGAL

BURKINA
FASO

ERITREA

DJIBOUTI

FEDERATED STATES
OF MICRONESIA

GAMB.
GUINEA-
BISSAU

GUINEA

BENIN

AFRICA

NIGERIA

CENTRAL
AFRICAN
REP.

ETHIOPIA

SRI
LANKA

CAMBODIA
BRUNEI
MALAYSIA

PALAU

KIRIBATI

SIERRA
LEONE

GHANA

LIBERIA

CÔTE
D'IVOIRE

TOGO

CAMEROON

SOMALIA

MALDIVES

Equator

NAURU

SAO TOME AND
PRINCIPE

EQUATORIAL
GUINEA

GABON

RWANDA

UGANDA

KENYA

I N D O N E S I A

EAST
TIMOR

PAPUA
NEW
GUINEA

SOLOMON
ISLANDS

CONGO

DEM. REP.
OF THE
CONGO

BURUNDI

TANZANIA

TUVALU

ATLANTIC
OCEAN

ANGOLA

ZAMBIA

MALAWI

SEYCHELLES

COMOROS

INDIAN
OCEAN

VANUATU

FIJI
ISLANDS

NAMIBIA

ZIMBABWE

BOTSWANA

MADAGASCAR

MAURITIUS

Tropic of Capricorn

AUSTRALIA

NEW
CALEDONIA
(France)

N
W   E
S

SOUTH
AFRICA

SWAZILAND

LESOTHO

MOZAMBIQUE

RÉUNION
(France)

AUSTRALIA

NEW
ZEALAND

FRENCH SOUTHERN &
ANTARCTIC LANDS
(France)

SOUTH GEORGIA &
SOUTH SANDWICH
ISLANDS (U.K.)

Antarctic Circle

ANTARCTICA

EUROPE

NORWAY

FINLAND

SWEDEN

RUSSIA

North
Sea

DENMARK

ESTONIA

Baltic Sea

LATVIA

LITHUANIA

RUSSIA

0   400 mi
0   400 km

IRELAND

UNITED
KINGDOM

NETH.

BELARUS

BELG.

GERMANY

POLAND

LUX.

UKRAINE

ATLANTIC
OCEAN

LIECH.

CZECH
REPUBLIC

SWITZ.

FRANCE

AUSTRIA

SLOVAKIA

HUNGARY

MOLDOVA

ROMANIA

SLOVENIA

MONACO

SAN MARINO

CROATIA

BOS. &
HERZ.

SERBIA

Black Sea

GEORGIA

ANDORRA

BULGARIA

PORTUGAL

SPAIN

ITALY

MONT.

MACEDONIA

ALBANIA

GREECE

TURKEY

GIBRALTAR (U.K.)

MALTA

MOROCCO

ALGERIA

TUNISIA

Mediterranean Sea

CYPRUS

SYRIA

LEBANON

GLOSSARY

atmosphere (AT-mus-feer) the blanket of gases that surrounds Earth

deforestation (dee-for-eh-STAY-shuhn) cutting down and burning of forests to clear land for farming and other development

emissions (ee-MISH-uhns) gases and pollutants sent into the air during combustion, which is the process of burning (such as the burning of gasoline in a car engine)

evaporation (ee-vap-oh-RAY-shun) to change a liquid, such as water, into a vapor or gas

fossil fuels (FOSS-uhl FEW-uhlz) combustible fuels, such as oil, gasoline, natural gas, and coal, which are drawn from deposits of organic material pressurized for millions of years

fossils (FOSS-uhlz) the ancient remains of a once-living thing, such as a skeleton

glaciers (GLAY-shurz) masses of ice slowly moving over the land

glider (GLY-dur) a plane that has no engine and stays aloft just by using air currents and airflow over its wings

greenhouse effect (GREEN-houss ih-FEKT) the trapping of the sun's heat in the atmosphere by greenhouse gases in the atmosphere

greenhouse gases (GREEN-houss GASS-ez) gases in the atmosphere, some of which may come from human activity, that trap outgoing radiation in Earth's atmosphere

phenomenon (fuh-NOM-uh-non) an event or a fact that can be seen or felt

solar radiation (SO-lur ray-dee-AY-shun) the energy given off by the sun

storm surge (STORM SURJ) the sudden rise of ocean water that occurs during strong winds in storms, such as hurricanes